Everything You Need to Know About *Genetically Modified Foods*

A researcher inspects the leaves of a papaya plant at the Monsanto Corporation's greenhouses. Scientists are working to genetically enhance the plants to make them resistant to the Southeast Asian strain of the papaya ringspot virus.

Everything You Need to Know About Genetically Modified Foods

Jeri Freedman

The Rosen Publishing Group, Inc.
New York

To my mother, Dolores Jardine, with love

Published in 2003 by The Rosen Publishing Group, Inc.
29 East 21st Street, New York, NY 10010

First Edition

Library of Congress Cataloging-in-Publication Data
Freedman, Jeri.
Everything you need to know about genetically modified foods/
by Jeri Freedman.
p. cm. — (The need to know library)
Summary: Details recent developments pertaining to genetic modification of food, including where the research is taking place and how it affects agriculture and the consumer.
Includes bibliographical references and index.
ISBN 0-8239-3612-0
1. Genetically modified foods—Juvenile literature. [1. Genetically modified foods. 2. Food—Biotechnology.]
I. Title. II. Series.
TP248.65.F66 F76 2002
641.3—dc21

2001008531

Manufactured in the United States of America

Contents

Introduction

According to the International Service for the Acquisition of Agri-biotech Applications (ISAAA), genetically engineered crops are grown on more than 100 million acres worldwide. The amount of land devoted to genetically modified crops was twenty-five times greater in 2000 than in 1996, and this trend is expected to continue. Genetically engineered produce (vegetables, fruits, and grains) is integrated into many types of food that are eaten throughout the world, including some food products we eat without even knowing they contain such ingredients. Genetic engineering has also been used to alter animals such as chickens, cows, sheep, and pigs.

Genetic engineering has been used to modify both plants and animals, such as chickens.

According to the True Food Network (at http://www.truefoodnow.org), genetically engineered ingredients have been found in popular foods such as Act II and Healthy Choice microwave popcorn, Quaker rice and corn cakes, Doritos corn chips, and Pringles potato chips.

Many companies in the food production industry see genetic engineering as a means of providing food that has better characteristics, such as enhanced longevity and increased nutritional value. Some organizations, such as the Food and Agriculture Organization (a branch of the United Nations) and the Center for Science in the Public Interest (a nonprofit group of scientists), see genetically engineered produce as a way to feed starving people in poorer parts of world. Others see the production of genetically modified foods as a way to reduce the amount of dangerous chemicals, such as pesticides and herbicides, that agriculture introduces into the environment.

There are, however, many dangers associated with genetically engineered produce. We need to be sure that genetically modified food is safe to eat and that it doesn't kill or change other plants or animals in the environment. It is likely that, over time, more and more of the food we eat will be genetically modified because the modifications being developed make crops easier to grow, result in produce that is easier to store and transport, and increase the amount of food produced by plants and animals. Therefore, it is important to understand both the positive and negative ways in which genetically modified food affects us.

Chapter 1

What Is Genetically Modified Food?

The term "genetically modified food" may bring to mind scenes from old science fiction movies of giant mutant tomatoes attacking people. The truth, however, is much simpler. "Genetic modification" means manipulating the traits, or characteristics, of plants or animals in order to produce specific features, such as the ability to withstand frost in a fruit or a lower fat content in cattle.

The New Foods

Characteristics of genetically modified produce may include better resistance to disease, tougher skins so that less damage occurs in shipping, and more vitamins. Livestock may be genetically modified to produce more meat. In addition, scientists have used genetic engineering technology to mass-produce compounds used in livestock to change the way their bodies work.

Farmers have practiced simple forms of genetic engineering for centuries, but for the past twenty years, scientists have focused their research on working directly with genetic structures.

For example, researchers have genetically engineered bacteria to produce bovine growth hormone (rBGH) in large quantities and have injected cows with this hormone to make them produce more milk. The average annual output of milk from a dairy cow is about 4,227 quarts (4,000 liters); cows treated with genetically engineered bovine growth hormone produce another 861 to 1,057 quarts (600 to 1,000 liters).

Genetically modified plants and animals are sometimes referred to as genetically modified organisms (GMOs). They are also sometimes called transgenic plants or animals. "Transgenic" means containing genes from outside the species.

Making the Best Plants and Animals

A simple type of genetic engineering has been practiced throughout the world for centuries. Farmers wishing to breed crops for characteristics like better disease resistance or improved flavor would grow plants that had those characteristics. They would then use the seeds from these plants to grow more the following year. Ranchers and dairy farmers wishing to produce larger cattle or cows that gave more milk would breed animals to have these characteristics.

Breeding plants or animals in the hope of obtaining the best offspring is an indirect method of improving a species because farmers do not directly change the genetic material that produces the organisms' characteristics. Beginning in the late 1980s, however, scientists started to change the genetic structure of plants and animals directly. They do this by inserting new genes into the cells of plants or animals that cause them to develop new characteristics. The genes used to alter plants can come from other types of plants or even from animals. The genes used to alter animals can come from the same kind of animal or from human beings.

Why Genetically Modify Organisms?

Directly changing the genes of plants and animals makes it possible for scientists to give organisms exactly the traits that people want. At the same time, it is possible for scientists to avoid accidentally giving a plant or animal a trait that they don't want it to have, which might happen if the organism was simply bred from other plants or animals.

Creating a Better Tomato

The most common type of genetic engineering is recombinant DNA technology. "Recombinant" means recombined or mixed. DNA is short for deoxyribonucleic acid, the material within a cell that tells the cell what substances or characteristics to produce. It is like computer software that provides instructions to a computer, telling it what to do. In recombinant DNA technology, scientists introduce DNA from one kind of cell into a different kind of cell and then use that cell to grow a plant or animal with specific characteristics. In the case of produce, the seeds of the new plant usually also carry the new type of DNA, and the plants grown from these seeds will also have the desired trait.

It is also possible to change more than one characteristic of a plant or animal through genetic modification. When more than one trait of an organism is genetically modified, it is called a stacked trait organism. Roundup Ready/YieldGard corn, manufactured by the Monsanto Company, is an example of a stacked trait plant. This type of corn is resistant to both insects and an herbicide (a compound that kills weeds) used on crops.

What Is a Gene?

The cells of all plants and animals consist of two parts: the outer part, which contains microscopic organs that carry out many activities (such as burning food for energy) necessary for keeping the body running, and the inner part—the nucleus—which is located in the center of the cell. The nucleus contains chromosomes, the elements that carry genetic information. That genetic information is coded into the DNA that makes up the chromosomes. A gene is one segment of a chromosome, similar to a line of code in a computer program. A gene (or sometimes a group of genes) determines a particular trait of a plant or animal. Adding, removing, or changing the gene for a particular trait will change that characteristic of a plant or animal.

How Are Plants Genetically Modified?

Plants can be grown directly from cells, from plant embryos (the early stage of a developing plant), or from tiny pieces of plant tissue that contain cells. In order to genetically modify plants, scientists must first obtain the gene they wish to insert into one of a plant's chromosomes. They can get the gene from another plant, an animal, or a bacterium. After scientists have isolated and removed the gene from its source, they insert it into the plant cells from which the new plant will grow.

Roundup Ready/YieldGard corn is an example of a stacked trait plant. It is resistant to both insects and an herbicide.

Two common techniques that scientists use to insert new genes into plant cells are the bacterial vector method and biolistics. In the bacterial vector method, scientists use a bacterium found in soil. This bacterium, called *Agrobacterium tumefaciens*, has developed the ability to insert DNA into plant cells to make the plants produce substances that enable the bacterium to live on them. Scientists insert a gene into the bacteria and then expose plant cells to the bacteria, which penetrate the plant cells. Once inside the cells, the genetically modified bacteria insert the gene they are carrying into the plant cells.

The second method is called biolistics. In biolistics, microscopic particles of gold or magnesium-tungsten are coated with a solution containing the genes to be transferred. Plant cells or plant embryos are placed in a sample dish, and a special device, sometimes referred to as a "gene gun," is used to shoot the particles into the cells. The particles penetrate the cells' nuclei, depositing the gene.

Tracking the New Gene

When scientists insert foreign DNA into a plant, there is no way to ensure that the new DNA will attach successfully. Therefore, the new gene is sometimes linked to, or combined with, a "marker gene." This marker allows the plants that have successfully taken up the new gene to be identified. For example, a gene that makes a plant produce vitamin A can be linked to a gene that makes a plant resistant to an antibiotic. After all of the plant cells have been exposed to the new gene, the whole batch of plant cells can be exposed to the antibiotic. Those cells that have successfully taken up the new gene—with its attached marker gene—will not be affected, since they are antibiotic resistant. However, cells that have not taken up the new gene will be killed by the antibiotic. Scientists can then grow plants from the cells that survived, knowing that those plants will contain the new gene.

How Are Animals Genetically Modified?

Microinjection is the technique commonly employed by scientists to genetically modify animals. In microinjection, the scientists remove fertilized egg cells from an animal and insert the new genes directly into the cells with a tiny syringe, hoping that the new genes will be incorporated into the animal's chromosomes. The eggs are then reinserted into a surrogate (or substitute) mother animal and are allowed to develop. Cloning is an example of an experimental technique that uses microinjection.

What Do the New Genes Do?

Most of the time, genes are inserted into a plant or animal in order to cause it to produce a specific chemical compound, such as a vitamin, or to increase its chances of survival.

One example of a plant that is being genetically modified to produce a specific compound is golden rice, which is being developed by Syngenta in Basel, Switzerland. This rice gets it name from its golden color, which is the result of its being genetically modified to produce betacarotene, or vitamin A. Vitamin A deficiency is a major problem in many developing nations, and it can cause blindness and even death. Golden rice can be grown in areas where the local diet lacks this vitamin, thereby helping people to avoid the negative effects of a vitamin A deficiency.

Alternately, genetic modification can be used to increase a plant's own chances of survival. For instance, a gene may be inserted into a plant that causes it to produce a substance that repels insects. This may make it possible to get bigger harvests from the plant and reduce the amount of pesticide needed to grow it. Similarly, a gene may be inserted into an animal to make it produce more growth hormone, causing it to grow more muscle and produce more meat. Some genetic modifications made to animals, however, can result in health problems. The disadvantages of genetically modifying animals are discussed in chapter 5.

Chapter 2

The History of Genetically Modified Food

In his work on the theory of evolution, English naturalist Charles Darwin (1809–1882) described the idea of natural selection. According to this theory, some animals and plants have traits that increase their ability to survive in a given environment. Over time, more of the plants or animals that have such traits survive and pass the traits on to the next generation. Thus, the traits become common in those types of plants or animals because of natural selection.

In contrast, when farmers deliberately choose to breed plants or animals with a specific trait instead of others, this is called artificial selection.

Charles Darwin developed the theory of natural selection, in which plants and animals with strong survival traits live to pass them on to subsequent generations.

Selective Breeding

The earliest form of genetic manipulation in plants and animals was selective breeding. In plants, this means picking out those with the most desirable traits, saving their seeds, and planting them in order to produce more plants with the desired trait. Similarly, farmers and ranchers would select animals with the most desirable traits and breed them. Such methods date back to the dawn of agriculture, which began in what is now Iran, Iraq, Israel, Turkey, and Southwestern Asia as early as 10,000 BC. Over the next several centuries, agriculture was developed in other regions, such as South America, and gradually spread throughout the world. The cultivation of wild species of plants was the earliest form of agriculture. When the plants were harvested, the seeds of the plants that had the best characteristics—such as the ability to produce the largest fruit— were saved for the following year's planting.

Traditional Methods of Improving Plants

Traditional methods of producing plants with specific traits include hybridization and cross-fertilization. Hybridization is the process of combining two types of plants to create a third type that has some of the characteristics of each of its "parents." For example, a nectarine is a hybrid of a peach and a plum. It has some of the characteristics of both of these fruits. Botanists (scientists who study plants) use hybridization to create completely new types of plants.

More commonly, however, botanists want to improve a type of plant that already exists. The most common method used to do this is cross-fertilization. In this method, pollen from one plant is introduced into the flower of another plant to produce more plants that have the traits of those specific parent plants.

Domestic varieties of plants are grown on farms and have been optimized through artificial selection. Many plants, such as corn, also exist in wild varieties

Gregor Mendel

In 1866, Gregor Mendel (1822–1884), a monk at a monastery in Brunn, Moravia (now Brnoin, part of the Czech Republic), established that specific traits of sweet peas could be passed from one generation to another. Mendel determined that traits were passed from generation to generation by means of what he called genetic factors (now called genes) and explained the principles that governed how the traits were passed on. Mendel's work laid the foundation for the science of genetics and the deliberate scientific breeding of plants and animals.

that have not been altered by artificial selection. If a domestic plant with a desired trait can't be found, they are sometimes cross-fertilized with one of their wild relatives to create a new plant with the desired trait.

Sometimes, however, the domestic and wild types of a plant can't be easily cross-fertilized. In these cases, breeders use other techniques, such as tissue culturing, in which new plants are grown from cells or little pieces of tissue taken from the parent plant. In addition, breeders sometimes expose the plant cells to various types of radiation, such as gamma rays (a type of high-energy radiation) or X rays, or treat them with chemicals, which causes their genes to mutate, or change. Scientists hope that this will result in new varieties of the plant with useful traits.

The Green Revolution

From 1965 to 1980, the use of artificial selection and scientific plant breeding, along with the use of fertilizers, resulted in a great increase in yields from plants like wheat and rice. This large increase in the productivity of food plants is referred to as the Green Revolution. However, by the mid-1980s, yields from food plants stopped getting larger, and it seemed as if it would be hard to get more improvement by traditional methods. These crops also needed large amounts of chemical fertilizers, pesticides, and water. This caused a great deal of concern about the effects on the environment from governmental agencies, such as the U.S. Environmental Protection Agency (EPA), and private activist groups, such as Greenpeace, as well as concerned citizens. The desire to create plants that need less artificial fertilizers, pesticides, and water was one of the reasons that scientists decided to try to genetically modify plants.

Early Genetic Modification of Plants

The first plant that was genetically modified by directly inserting a gene into it was a tobacco plant. It was modified in 1983 to make it resistant to an antibiotic. In 1993, the first genetically modified food plant, the Flavr Savr tomato, was created by Calgene, Inc., in Davis, California. The Flavr Savr was a delayed-ripening tomato; because it took longer to ripen, it remained firmer during transportation to supermarkets. In the past decade, research has been undertaken on the development of genetically modified apples, melons, broccoli, corn, grapes, peanuts, potatoes, rice, and other produce.

Early Genetic Modification of Animals

The first animal to be successfully modified genetically was a mouse developed for use in cancer research in 1987 at Harvard University. In the years since, forty different types of animals have been genetically modified, including pigs, fish, and cows. In 1997, genetic technology was extended to cloning. In cloning, the nucleus is removed from an egg cell of an animal. The nucleus of a cell from the animal to be cloned is inserted in place of the removed nucleus. The egg is then inserted into a surrogate mother and

 Dolly, the first animal to be genetically cloned from adult sheep cells, was named for the singer Dolly Parton.

allowed to develop. The first animal to be produced by this technology was a sheep named Dolly, who was cloned in 1997 by Dr. Ian Wilmut at the Roslin Institute in Scotland.

Many companies that produce genetically modified produce, as well as government and private advocacy groups that support the genetic modification of food plants and animals, see it as an advanced type of selective breeding and artificial selection. Many activist and consumer organizations who oppose the practice, however, see it as unnatural and worry about the dangers of inserting foreign genes into a plant or animal.

Chapter 3

The Advantages of Genetically Modified Food

Throughout the world, genetic engineering is being used to produce crops that are resistant to insects and disease. For example, because bananas are a major part of the diet of many people living in developing countries, such as Uganda and other African nations, an international team of scientists is currently working to develop a new type of banana. According to the Web site for the University of Leeds School of Biology (at http://www.biology.leeds.ac.uk), in 2000, 85 million tons of bananas were grown worldwide. Most bananas are eaten in the country where they are grown, and people depend on them as a major food crop. However, they are vulnerable to insects, such as the nematode (a parasitic worm), and to viruses that kill plants. Scientists are examining the gene of a species of South Asian banana, which they plan to genetically modify into a variety that is more resistant to these pests and diseases.

Improving Livestock

An important goal of research on livestock is to develop animals that are resistant to common diseases, such as hoof and mouth disease in cattle. Work is also being done by numerous companies and academic research organizations in North America and Europe to develop animals that can grow more muscle. This would result in more meat per animal and animals whose meat has a lower fat content, which may be of interest to people who are concerned about their health.

Eliminating Hunger

In 2000, the population of the world was approximately 6 billion. According to the International Service for the Acquisition of Agri-biotech Applications (ISAAA), it is expected to reach 9 billion by 2050. By that time, the ISAAA expects that 90 percent of the population will live in Asia, Africa, and Latin America. It further estimates that 840 million people in developing nations suffer from malnutrition, a condition in which people become ill because they don't get enough nutrients.

One of the greatest challenges facing us is the ability to produce enough food to feed everybody. What makes this doubly challenging is that natural resources are becoming less available as water and nutrients in the soil are being used up faster than they can be replenished by nature. Some rain forests have been destroyed to make

Cattle are vaccinated for hoof and mouth disease, a contagious illness. Genetic engineering could make animals resistant to common diseases.

room for farmland. However, this is not ecologically desirable because the trees in the rain forest are an important source of the oxygen we need to survive, and the rain forest itself provides a home for many unique and useful plants and animals. In addition, the world faces a potential water shortage because of the huge quantities used by an ever-increasing population for personal consumption, agricultural irrigation, and industrial purposes such as mining and manufacturing operations.

To make matters worse, much of the available farmland in developing nations and poor rural areas has been depleted of nutrients (such as nitrogen, potassium, and phosphorus) needed for plant growth or is of poor quality for growing food. Organizations like the United Nations Development Programme, which is in favor of genetically modified produce, believe that these crops have the potential to provide food for local people. For example, a team of scientists led by Dr. Eduardo Blumwald at the University of California at Davis has produced a tomato that will grow in soil too salty for normal vegetables to grow. Other plants are being developed that require less water to grow, which will allow crops to be grown in arid regions such as the Middle East.

Developing Better Plants

The following are some ways in which genetic engineering could produce better vegetables that would improve nutrition and/or feed more people.

More Productive Plants

Dwarf plants usually produce more fruit and less leaves and stalks. Such dwarf genes are commonly found in plants like wheat and rice. If dwarf genes from these plants were inserted into other types of vegetables, this might result in higher-yielding varieties. In addition, some varieties of plants that are under development have increased photosynthesis, the process by which plants use sunlight to nourish themselves. Increased photosynthesis leads to greater growth and more produce from plants.

Hardier Plants

Plants could be developed that tolerate extreme temperatures, floods, and/or droughts. For example, scientists in India are working on a variety of rice that can survive submersion in floodwaters for a long period of time. In addition, plants could be developed that grow in areas with short growing seasons.

Plants That Can Help Conserve the Environment

Presently, agriculture requires soil that meets very specific conditions. It can't be too salty or too acidic, or contain toxic elements such as aluminum that inhibit plant growth. In many developing tropical and subtropical nations, there isn't enough cleared land that meets these conditions, and forests or jungles must be cut down in order to grow food. Plants could be developed that can grow in soil that is not good for growing traditional plants. Scientists in Mexico, for example, are working on a variety of corn that would be aluminum resistant.

Disease-Resistant Plants

Many crops, such as wheat, rice, millet, sweet potatoes, and bananas, are subject to diseases that can kill them. Genetic modification could be used to develop varieties of plants that are resistant to common plant diseases, such as the cucumber mosaic virus.

Longer-Lasting Vegetables

Slower maturing varieties of vegetables could be developed that would last longer. This way, less produce would be wasted, and it could be stored for longer periods of time in areas that lack electricity for refrigeration.

More Nutritious Vegetables and Grains

There are millions of people in developing nations who suffer from nutritional deficiencies (a lower than necessary amount of one or more important nutrients). We've previously discussed plants that produce vitamins. Plants could also be developed to provide other nutritional benefits. For example, plants grown to make vegetable oils and margarine could be genetically engineered to produce healthier oil. Calgene, Inc., is developing a type of canola that produces oil that contains less of the bad type of fat that clogs up blood vessels. Plants could also be developed to produce more protein and provide better nutrition in areas where food is scarce.

Chapter 4

The Politics of Food

Growing and eating genetically modified produce raises many issues related to the economic and social effects on people. For instance, sometimes genes from animals are inserted into plants. An example of this is a gene from a fish that lives in the Arctic that has been inserted into a variety of tomato to make it more resistant to frost. This allows farmers to have a longer growing season for tomatoes. This is especially beneficial in northern climates, which have short growing seasons. Incorporating animal genes into plant produce has sparked debate among activists, representatives of industry organizations, and legislators in countries such as India.

India has a population of approximately one billion people. Faced with the need to feed this extremely large population, the government is very interested in genetically modified plants. However, the majority of India's people observe the Hindu religion, which requires its followers to be vegetarian. Many vegetarians are worried about the possibility of non-vegetarian elements being introduced into their food. This concern has resulted in opposition to genetically modified food from activists in India.

What Should Be Grown in Developing Countries?

Questions that have been raised by activist groups and organizations concerned with improving the standard of living in developing nations include:

◎ **Is genetic engineering necessary to produce enough food in developing nations?**

◎ **What type of produce should people be encouraged to grow?**

◎ **Will genetically engineered produce make people more independent or more dependent on developed nations for their food?**

◎ **Will genetically engineered food violate religious taboos of some populations?**

Objections to Genetically Modified Vegetables

The problem of feeding those who are starving is not a matter of producing adequate food, but rather of distributing the food so that more people can be fed. According to this argument, enough food is produced worldwide to feed every person on the earth. However, there's an excess of this food in the United States and Canada, while at the same time, the food is not distributed to people in developing nations. The problem with relying on redistribution is that someone—the farmers or governments of developed nations, or the governments of the developing nations—must pay for this food and the costs associated with its distribution. The farmers and governments of developed countries are not willing to foot the bill, and the developing nations don't have the money to buy food.

Some of the present problems are the result of the policies of organizations such as the World Trade Organization (WTO), which focus on encouraging trade among nations. The WTO, for example, encourages farmers to grow crops to make money. This leads them to devote large amounts of available farmland to crops that can be sold and are widely eaten around the world, such as white rice, but which have little nutritional value. It might provide a healthier diet for the local population if the farmland was divided into smaller parcels of land and used instead to grow a wider variety of food crops, such as different types of vegetables, that were sold locally.

Patenting Life

Another issue is that new varieties of plants developed by genetic engineering techniques can be patented, meaning that only the company that developed them or organizations that pay royalties to that company can sell them. In December 2001, the U.S. Supreme Court upheld the right of seed companies, such as Pioneer, a subsidiary of DuPont, to prohibit other companies or farmers from selling or reusing their seed without paying for the right to do so. This means that farmers who want to replant the seed will have to pay a fee to the patent-holding company, which makes it more costly for farmers to grow genetically improved crops. The desire to maintain control over patented seeds has led some companies to attempt to develop "terminator technology," in which the seeds of the plants die or will not grow new plants.

Historically, farmers have harvested seeds from one year's crop to grow the next year. With these new varieties of plants, it would be necessary for farmers to purchase new seeds from the company every year. If one of the major justifications for the development of genetically engineered foods is to feed people in developing nations, then this technology is at odds with this goal.

The policies of the World Trade Organization in Geneva, Switzerland, have encouraged trade but compromised the distribution of food to people in poor countries.

Not all experts are worried about the bad effects that growing genetically modified foods may have on developing nations. Instead, some experts feel that not enough is being done to get this technology into developing countries where it is most needed. The United Nations Development Programme (UNDP) has expressed concern that the objections of people in the developed nations of North America and Europe who are worried about the safety of genetically engineered foods may be keeping these crops from being used by those in developing nations. Twenty-one countries in sub-Saharan Africa, including Ghana, Nambia, and Nigeria, are facing the problems of hunger or famine. The UNDP suggests that developing nations should be allowed to do their own evaluations of the risks associated with genetically engineered produce such as drought-resistant corn.

Finally, in order for the benefits of genetically modified produce to be seen, commercial developers of these new technologies must work with governments and nonprofit organizations to ensure that such crops are available in the areas where they are most needed. Also, local populations must be trained in the planting and maintenance of these new foods. This requires the development of a local administrative structure to support the cultivation of the genetically modified crops. The training and administrative development could be handled by the United Nations through its Food and Agriculture Organization, United Nations Development Programme, or other divisions. The process also requires the involvement of the governments of the countries where this type of agriculture is being implemented, and it may require the cooperation of companies that have technical expertise in the growth of genetically modified crops.

Chapter 5

Concerns About Genetically Engineered Food

The debate over whether genetically modified foods are damaging to the environment or dangerous to our health has been carried out through protests and demonstrations, articles in newspapers and magazines, and legislative efforts. Some people ask if we should be tampering with our food supply at all. What if we create a species that goes wrong and actually damages our ability to produce food? Others feel that genetic engineering may result in our growing only a few preferred varieties of a given crop, and varieties that occur in nature may die out. Do we have the right to alter the environment, and do we really understand what the effects of doing so might be?

Effects on the Environment

One concern about planting genetically modified vegetables is that they might spread their genes to native plants, resulting in unintended changes to other plant species. This could be a problem if a gene for herbicide resistance were to spread from corn to weeds, for example. To date, there has been little evidence of such gene transfers taking place between different species of plants; however, we can't be certain that such transfers won't occur if genetically engineered plants were to be widely cultivated.

A related concern is the spreading of pollen from a genetically engineered variety of crop to a closely related non–genetically modified variety grown close by. There have been instances, for example, of pollen from test crops of genetically modified corn spreading to nearby fields of non–genetically modified corn.

Effects on Beneficial Insects

Another concern about the effect of genetically modified crops on the environment relates to the plants' potential to harm native animals or insects. Much concern has been expressed over a variety of corn that was developed by the Ciba-Geigy Corporation (now Novartis) and Monsanto and its effect on the monarch butterfly. The Bt gene—which makes corn resistant to corn-destroying insects—was used in making this variety of corn. In

Demonstrators march in the streets of Montreal, Canada, to protest genetically modified organisms.

1992, researchers at Cornell University performed a study in which they demonstrated that when monarch butterfly larvae were fed large amounts of Bt corn pollen, about half of them died. The study was performed in a laboratory and did not necessarily reflect how the insects would react in the wild, but it raised valid concerns about the potential effects that genetically modified plants might have on beneficial insects.

Another issue associated with the development of pesticide-producing plants like Bt corn is that in the past, whenever a new pesticide has been introduced, insects have developed resistance to that chemical over time. This happens because those members of the insect population that are susceptible die off and those who are resistant to the chemical survive and breed. Therefore, it is possible that large numbers of insects exposed to Bt corn over time will become resistant to the naturally produced pesticide and eliminate its effectiveness.

Effects on Biodiversity

Biodiversity refers to the number of different types of plants and animals in the environment. Maintaining biodiversity may be important in ensuring that we have types of plants and animals that will survive under different conditions. In the past several decades, the biodiversity of food plants has decreased. As new, higher-yielding varieties of food crops have been developed by conventional methods, more and more farmers in the United States, Canada, and other countries have chosen to grow those varieties. Some people fear that the introduction of genetically modified varieties of vegetables—with more desirable traits such as drought, disease, or insect resistance—will make this problem worse.

Animal Welfare

Genetically modifying animals such as cows, chickens, and fish in order to produce more growth hormone results in larger animals and more meat. However, this excessive growth can cause painful and crippling health problems for the animals. In addition, when human genes are used to genetically modify animals so that they will produce compounds useful for human health, the animals can develop joint, eye, and other serious health problems. If animals are genetically modified, it must be done in a way that does not cause the animals unnecessary suffering.

Effects on Humans: Are the Genes Safe to Eat?

Certain genetic engineering techniques use a gene that gives plants resistance to specific antibiotics as part of the genetic engineering process. Some people are concerned that a gene producing antibiotic resistance might be transferred to human beings in large amounts through the foods that we eat. If this happened and germs in our environment were exposed to large numbers of people with this antibiotic resistance, the result could be an increase in antibiotic-resistant germs, making it harder to fight some diseases. For this reason, researchers are working on developing new techniques for creating genetically engineered plants that don't require the use of antibiotic-resistant genes.

New Chemicals in Our Food

The safety of compounds produced by genes inserted into genetically modified vegetables is also a cause for concern. Pesticides that are sprayed on vegetables can

The meat of hybrid turkeys like these may contain traces of chemical compounds and hormones that can be transferred to people who eat it.

be washed off before food is consumed. However, when such compounds are produced by the plant itself—as is the case with Bt corn—people will be eating these compounds when they eat the vegetables. At present, we do not know about the long-term health effects of eating significant amounts of such compounds.

A related problem exists with meat from genetically engineered animals. When animals are genetically engineered to produce increased amounts of growth hormone or other compounds, traces of the compounds remain in the meat that people eat. We do not yet fully understand what effects these compounds may have on the people who eat the meat.

Allergies and Genetically Modified Foods

One concern with genetically engineered food is that some people might be allergic to compounds produced by new genes added to a familiar animal or vegetable. Some allergic people respond to such exposure with mild symptoms, such as sneezing or runny eyes. Other people, however, can have serious or even fatal responses.

Those in favor of genetically engineered plants claim that even with traditional cross-fertilization and hybridization methods, new substances may be introduced into the modified strain of a vegetable. Those who object to genetically modifying foods feel that the ability to introduce material from other plants, and even animals, will increase the range of allergens (allergy-causing substances) that might be introduced, without any awareness on the part of the consumer.

For instance, many people are allergic to peanuts. What would happen if a gene from the peanut were introduced into another food crop, making it produce an allergy-causing substance produced by peanuts? Would people who are allergic to peanuts have an allergic reaction to it? How would people who are allergic to peanuts know that eating such a food might be dangerous for them? How might we address the problem of introducing potential allergens into vegetables?

Two useful approaches are testing and labeling. Because a known gene is used to express a specific substance, it is possible to test whether exposure to that substance is likely to cause an allergic reaction. Returning to the peanut example, if a peanut gene is inserted into a vegetable and causes it to produce a substance commonly produced by peanuts, it is possible to use both laboratory tests and tests on people to identify whether or not those with peanut allergies will have a reaction to the genetically modified vegetable. While this won't guarantee that no one will be allergic to the plant, it would provide an indication of whether or not there is a large risk. Safety issues relating to genetically modified foods are currently under consideration in the United States, Canada, and other countries around the world. These issues, as well as the labeling of genetically modified food, are discussed in the next chapter.

Chapter 6

Coming Soon to a Supermarket Near You

To date, very few genetically modified vegetables or animal products have been approved for people to eat. However, in the future more and more such foods are likely to appear in supermarkets and as ingredients in prepared foods.

Identifying Genetically Modified Food

Unless such foods are labeled, people will not know that the food they are eating has been genetically modified. The labeling of genetically engineered food, however, has become a controversial topic. Many food growers and packagers of processed food object to labeling because they fear the public will be afraid to buy food that is genetically modified. On the other hand, consumers who are concerned about the safety of such foods feel that they have the right to decide for themselves whether they want to eat it. In addition, those who have allergies to specific substances found in nuts, fish, shellfish, or other plants and animals need to know what foreign substances genetically modified vegetables might contain.

Labeling Issues

Currently, the U.S. Food and Drug Administration (FDA) requires that food labels contain information that is "material," or significant. Nonetheless, there is no consensus among producers of genetically modified foods, consumers, and activists on what information is considered to be significant.

In an online interview, Dr. James Maryanski, the Biotechnology Coordinator in the FDA's Center for Food Safety and Applied Nutrition, indicated that the FDA would likely require labeling of genetically engineered foods containing genes from plants or animals that commonly cause allergic reactions, such as peanuts or fish, unless the company developing the food can show scientifically that the food doesn't cause allergic reactions. He also said that labeling might be required if the nutritional content of food was affected by genetic engineering.

One problem with attempting to label genetically engineered fruits or vegetables is the difficulty of identifying them when they are mixed with non–genetically modified fruits or vegetables in canned goods. This

There is strong consumer demand for "bio label foods," which guarantee that they are free of chemicals and hormones, and meat from livestock that is raised in a humane fashion.

might happen, for example, if genetically modified apples were mixed with conventional apples to make applesauce at a processing plant.

This issue becomes even more complex when genetically modified produce is processed before being used to manufacture prepared foods. For example, genetically modified soybeans may be processed in powder form, which in turn may be sold to food manufacturers who use it as a thickening agent. Tracking genetically engineered produce from the whole vegetable stage through various processing steps to the final product makes the prepared food manufacturing process more complicated and expensive.

Monitoring Genetically Modified Foods for Safety

Many countries are taking steps to regulate genetically modified food.

In Canada

In Canada, the regulation of genetically modified food is handled by the Canadian Food Inspection Agency (CFIA), Health Canada, and Environment Canada. Health Canada is responsible for ensuring that genetically modified foods do not have negative effects on people's health and for labeling related to possible health risks. Meanwhile, the CFIA is responsible for general labeling and advertising regulations, and Environment Canada is responsible for ensuring that genetically modified plants and animals are safe for the environment.

At present, labeling genetically modified food in Canada is voluntary, and approval to market genetically modified food is given on a case-by-case basis. The Canadian Biotechnology Advisory Committee (CBAC) has called for the agencies responsible for the regulation of genetically modified food to develop a standard set of tests for use in approving genetically modified food. They also require the mandatory labeling of genetically modified food or the availability of a central source where consumers can obtain information on which products contain genetically modified ingredients.

In the United States

In the United States, the FDA, the EPA, and the U.S. Department of Agriculture (USDA) share joint responsibility for regulating genetically engineered food. The USDA is responsible for overseeing genetically modified plants and animals and their distribution. The EPA has control over pesticides and pesticidal agents, and regulates genetically engineered plants that incorporate genes that produce natural pesticides. The FDA is responsible for ensuring that food additives and food products are safe to eat. They approve genetically modified food for sale in supermarkets and for use in prepared food.

One way to ensure that genetically modified foods are safe to eat is to require testing of all types of genetically modified food prior to their release. Such testing should show that the foods do not produce allergic reactions or other health problems.

The EPA has blocked the release of some genetically engineered products that it felt had not been tested enough to show that they are safe for human consumption. For example, in July 2001, the EPA refused to allow the use of StarLink corn, developed by Aventis CropSciences, to be used as an ingredient in prepared foods for human consumption. This decision was made because the EPA felt that not enough testing had been done to show that the genetically modified corn would not cause allergic reactions in consumers.

Currently, the FDA does not require safety testing for most genetically engineered food. In 1992, the FDA issued a policy stating that genetically engineered foods would be considered to be substantially equivalent to (the same as) conventional varieties of food unless evidence showed otherwise.

The Center for Food Safety in Washington, D.C., has indicated that in March 2000, a group of scientific, consumer, environmental, and farm organizations filed a petition with the FDA calling for it to develop a standard set of tests for genetically engineered foods and demanding the labeling of genetically engineered foods. The petition called on the FDA to do the following:

◎ **Replace its 1992 policy on genetically modified foods with one requiring testing and labeling of such produce**

◎ **Require specific testing to see if the food produced allergic reactions**

◎ **Require an environmental analysis for each genetically engineered food to see if it is safe to grow**

◎ **Require genetically engineered foods to be labeled**

In response to the widespread scientific and public concern about the need to ensure that genetically modified foods are safe to eat, the FDA is reviewing its 1992 policy, although it is not yet clear what changes (if any) the FDA will make.

In 1961, the Codex Alimentarius Commission, which currently has 165 members, was formed jointly by the World Health Organization (WHO) and the Food and Agriculture Organization (FAO). In July 2000, the commission issued a set of universal principles relating to the safety of genetically modified foods. Basically, the principles state that governments should test foods made from genetically engineered produce to make sure they are safe to eat (especially in terms of potential allergic reactions) prior to releasing them to the general public. The commission has also called for labeling of any genetically modified foods that are shown to cause allergic reactions. The Ad Hoc Intergovernmental Task Force on Foods Derived from Biotechnology, established by the Codex Alimentarius Commission in 1999, is attempting to develop standards for genetically modified foods by a target date of 2003. Such standards could provide a consistent way of ensuring the safety of genetically modified foods.

What's That in My Taco?

In the future, food may provide more than nutrition. One area of genetic modification being explored is the development of foods that contain vaccines or medications. Research is being done on growing plants that might produce vaccines for diseases such as hepatitis B and cholera. Growing varieties of vegetables that contain oral vaccines would allow such vaccines to be distributed in remote areas and developing countries where conventional vaccines cannot be distributed because there aren't adequate storage and distribution facilities. Such genetically modified plants could provide such areas with a readily available, inexpensive supply of vaccines. This method could also be used to have plants produce other types of medications.

In recent years, there has been a great deal of interest in dietary supplements (substances eaten to improve health). Genetic modification can be used to create produce that has higher levels of substances such as antioxidants (compounds that fight damage to cells) that would appeal to those interested in health and fitness. Fruits and vegetables could also be genetically engineered to produce vitamins. However, such attempts must be undertaken with caution because eating too much of some vitamins can cause health problems. Therefore, while vegetables that produce vitamins may be good for people who do not otherwise get enough in their diets, eating too much of such fruits or vegetables may be bad for people who are already getting enough of a given vitamin.

Research is also being done on genetically modifying animals to produce compounds that could be used to treat human diseases such as cancer, cystic fibrosis, and diabetes, among others. In some cases, scientists are attempting to genetically engineer cows so they will produce medically beneficial compounds in their milk.

Another area that is being investigated is producing chicken eggs that contain compounds that could be used to improve people's health or treat disease. When animals are raised to produce medically beneficial compounds, this is sometimes called pharming. When dietary supplements or medical compounds are grown in genetically modified produce or livestock, they are called nutraceuticals. Finally, scientists are working to develop nonfood plants, such as cotton, that produce synthetic (artificial) fibers and plastics.

It is very likely that as time goes by, more and more genetically modified food products will appear in what we eat. It is important that we approach raising and eating such produce and livestock carefully to protect both ourselves and our environment.

Glossary

allergen An allergy-causing substance.

artificial selection The deliberate breeding of selected plants so that desirable characteristics will be passed on to their offspring.

bacterial vector A bacterium that transmits a gene to the nucleus of a cell.

betacarotene Vitamin A.

biodiversity The number of different types of plants and animals in an environment.

biolistics The method of inserting genes into a plant cell by attaching them to metal particles and "shooting" them into the cell.

botanist A scientist who studies plants.

cloning A process in which the nucleus is removed from an egg and replaced by the nucleus of a cell taken from a parent animal. The egg is reinserted into a surrogate animal and allowed to develop.

Codex Alimentarius Commission A committee created jointly by the World Health Organization and the Food and Agriculture Organization; it is working to establish regulations for genetically modified food.

cross-fertilization The introduction of pollen from one plant into the flower of another to breed plants with specific traits.

DNA (deoxyribonucleic acid) The material that makes up genes and carries the information that tells cells what substances to produce.

embryo The early stage of a plant's or animal's growth.

genetically modified organism (GMO) A plant or animal whose DNA has been directly altered.

genetic engineering The process of directly changing the genes of plants or animals to produce specific characteristics.

herbicide A chemical compound that kills weeds.

hormone A chemical produced in the body that affects the way the body functions.

hybridization The process of combining two types of plants to create a third type of plant that has some of the characteristics of each of its "parents."

marker gene A gene attached to another gene being inserted into a plant cell in order to allow scientists to identify those plants that have incorporated the desired gene.

microinjection The technique by which genes are directly inserted into the nucleus of a cell via a tiny syringe.

mutate To change; this sometimes happens to the genes of a plant when they are treated with radiation or chemicals.

natural selection A process in which plants with the best traits survive and pass their characteristics along to their offspring.

nucleus The center of a cell in which the components that contain genes are found.

nutraceuticals Medically beneficial compounds produced in genetically modified plants or animals.

pesticide A chemical compound that kills insects.

pharming The raising of genetically modified animals to produce medically beneficial compounds.

photosynthesis The process by which plants use sunlight to nourish themselves.

produce Vegetables, fruits, and grains.

protein A basic compound that makes up plant or animal tissue.

stacked trait plant A plant in which more than one characteristic has been genetically modified.

surrogate A substitute; refers to an animal into which a genetically modified egg is inserted and is allowed to develop.

For More Information

In the United States

The Center for Food Safety
660 Pennsylvania Avenue SE, Suite 302
Washington, DC 20003
Web site: http://www.centerforfoodsafety.org

Monsanto Company
800 North Lindbergh Boulevard
St. Louis, MO 63167
(314) 694-1000
Web site: http://www.monsanto.com

Union of Concerned Scientists
2 Brattle Square
Cambridge, MA 02238
(617) 547-5552
Web site: http://www.ucsusa.org/agriculture/index.html

U.S. Food and Drug Administration
Center for Food Safety & Applied Nutrition
5100 C Paint Branch Parkway
College Park, MD 20740-3835
Web site: http://www.cfsan.fda.gov

In Canada

Canadian Biotechnology Advisory Committee
240 Sparks Street, 05 West
Ottawa, ON K1A 0H5
(866) 748-2222
Web site: http://www.cbac-cccb.ca

Canadian Food Inspection Agency
59 Camelot Drive
Ottawa, ON K1A 0Y9
Web site: http://www.inspection.gc.ca

Food Biotechnology Communications Network
1 Stone Road West
Guelph, ON N1G 4Y2
(519) 826-3440
Web site: http://www.foodbiotech.org

Web Sites

Due to the changing nature of Internet links, the Rosen Publishing Group, Inc., has developed an online list of Web sites related to the subject of this book. This site is updated regularly. Please use this link to access the list:
http://www.rosenlinks.com/ntk/gemf/

For Further Reading

Anderson, Luke. *Genetic Engineering, Food, and Our Environment.* White River Junction, VT: Chelsea Green Publishing, 1999.

Cummins, Ronnie, and Ben Lilliston. *Genetically Engineered Food: A Self-Defense Guide for Consumers.* Berkeley, CA: Marlowe & Company, 2000.

Manning, Richard. *Food's Frontier: The Next Green Revolution.* New York: Farrar Straus & Giroux, 2000.

Nottingham, Stephen. *Eat Your Genes: How Genetically Modified Food Is Entering Our Diet.* New York: University of Capetown Press, 1999.

Teitel, Martin, and Kimberly A. Wilson. *Genetically Engineered Food: Changing the Nature of Nature.* Rochester, VT: Park Street Press, 2001.

Tokar, Brian, ed. *Redesigning Life? The Worldwide Challenge to Genetic Engineering.* New York: Zed Publishing, 2001.

Bibliography

Agriculture Department, Food and Agriculture
    Organization of the United Nations (FAO).
    "Biotechnology in Agriculture." January 1999.
    Retrieved August 15, 2001 (http://www.fao.org/ag/
    magazine/9901sp1.htm).

Anderson, Luke. *Genetic Engineering, Food, and Our
    Environment.* White River Junction, VT: Chelsea
    Green Publishing, 1999.

BBC News. "Genetically-modified Q&A." April 6, 1999.
    Retrieved August 15, 2001 (http://news.bbc.co.uk/
    hi/english/special_report/1999/02/99/
    food_under_the_microscope/newsid_280000/
    280868.stm).

Environment News Service. "UN Calls for GM Food
    Safety." Lycos Wired News. July 10, 2001. Retrieved
    August 15, 2001 (http://www.wired.com/
    news technology/0,1282,45108,00.html).

Fox, Michael W. *Beyond Evolution: The Genetically Altered Future of Plants, Animals, the Earth . . . and Humans.* New York: The Lyon Press, 1999.

Kaufman, Marc. "EPA Rejects Biotech Corn as Human Food." *Washington Post* Online. July 28, 2001. Retrieved August 15, 2001 (http://www. washingtonpost.com/wp-dyn/articles/ A62091-2001Jul27.html).

Keep Nature Natural. "The Making of Genetically Engineered Foods." Retrieved August 15, 2001 (http://www.keepnatural.org/science.html).

Keep Nature Natural. "Policy & Politics." Retrieved August 15, 2001 (http://www.keepnatural.org/ policypolitics.html).

Keep Nature Natural. "Potential Negatives of GE Foods." Retrieved August 15, 2001 (http://www. keepnatural.org/risksbenefits.html).

Mandel, Charles. "Hidden Wheat Fields Spark Outrage." Lycos Wired News. August 7, 2001. Retrieved August 15, 2001 (http//www.wired.com/news/ medtech/0,1286,45811,00.html).

Monsanto. "Achievements in Plant Biotechnology 2000." Retrieved August 15, 2001 (http://www.monsanto. com/monsanto/about_us/environmental_ information/plant_biotech01/biotech.pdf).

Nottingham, Stephen. *Eat Your Genes: How Genetically Modified Food Is Entering Our Diet.* New York: University of Capetown Press, 1999.

O'Connor, Anahad. "Altered Tomato Thrives in Salty Soil." The *New York Times* on the Web. August 14, 2001. Retrieved August 15, 2001 (http://www. nytimes.com/2001/08/14/science/life/14TOMA.html).

Philipkoski, Kristen. "Protestors Steamed Over Rice." Lycos Wired News. June 26, 2001. Retrieved August 15, 2001 (http://www.wired.com/news/medtech/ 0,1286,44804,00.html). The

Pro Gobal. "Dispelling 'Frankenfear.'" July 17, 2000. Retrieved August 15, 2001 (http://www.geocities. com/socialism_2000/pages/genetic.html).

Reaves, Jessica. "Are First World Fears Causing the Third World to Go Hungry?" Time.com. July 9, 2001. Retrieved August 15, 2001 (http://www.time. com/time/world/article/0,2599,166925,00.html).

Ticciati, Laura, and Robin Ticciati. *Genetically Engineered Foods: Are They Safe? You Decide.* Los Angeles, CA: Keats Publishing, 1998.

True Food Network. "True Food Shopping List: Snack Foods." Retrieved October 26, 2001 (http://www. truefoodnow.org/gmo_facts/product_list/snack.html).

Index

About the Author

Jeri Freedman has a B.A. from Harvard University and spent fifteen years working in companies in the biomedical and high-technology fields. She is the author of several plays and, under the name Foxxe, is the coauthor of two science fiction novels. She lives in Boston.

Photo Credits

Cover © Sonda Dawes/The Image Works; pp. 2, 24, 27, 35, 42, 47 © AP/Wide World Photos; p. 7 © Grant Heilman Photography Inc./Index Stock; pp. 10, 14 © Monsanto Company; p. 19 © Ewing Galloway/Index Stock; p. 39 © Reuters New Media Inc./Corbis.

Book Design
Tom Forget

Layout
Tahara Hasan